Math Critical Thinking

Creative Puzzles to Challenge the Brain

Pamela Amick Klawitter, Ed.D.

Illustrated by Bev Armstrong

The Learning Works

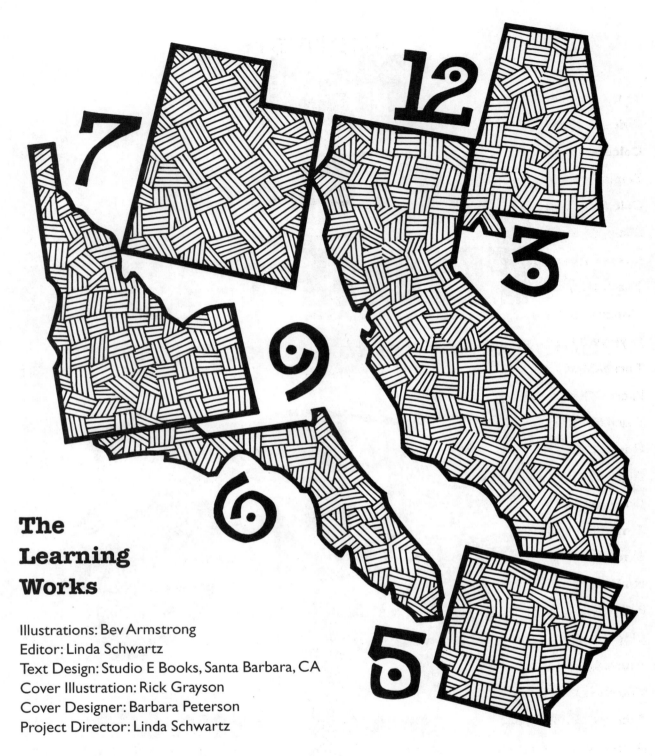

The Learning Works

Illustrations: Bev Armstrong
Editor: Linda Schwartz
Text Design: Studio E Books, Santa Barbara, CA
Cover Illustration: Rick Grayson
Cover Designer: Barbara Peterson
Project Director: Linda Schwartz

Contents

To the Teacher

Math Critical Thinking gives students the opportunity to develop critical thinking and problem-solving skills. Activities can be completed by individual students, or students may challenge one another. Skills include the following:

- convergent and divergent thinking

- brainstorming

- categorizing/classifying

- visual memory

- ideational fluency

- vocabulary development

- flexibility

- force-fitting/forced association

- originality

- forecasting

- synthesizing

Twice as Nice

Use the two-digit numbers shown below to answer the following questions:

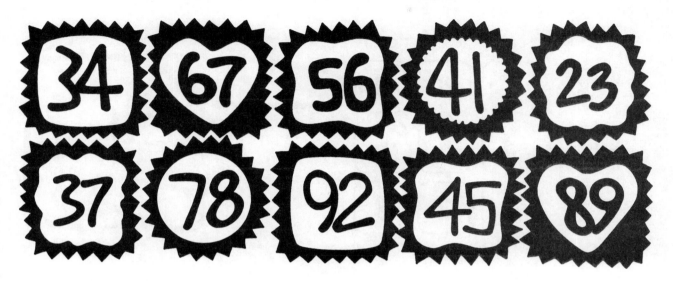

_____ 1. Which two numbers added together give the sum closest to 100 without going over?

_____ 2. Which two numbers multiplied together give the product closest to 1,500 without going over?

_____ 3. What is the sum of all the numbers that are evenly divisible by 3?

_____ 4. Use pairs of the two-digit numbers above to create four-digit numbers. What is the sum of the three smallest four-digit numbers you can create?

_____ 5. What is the sum of the two greatest six-digit numbers you can create using three pairs of the two-digit numbers above?

_____ 6. What number would need to be added to the list to make the sum of the even numbers equal to the sum of the odd numbers?

_____ 7. What is the sum of numbers whose first and second digits are consecutive numbers?

_____ 8. Use three pairs of the two-digit numbers in the list to create six-digit numbers. What is the difference between the largest and smallest six-digit numbers you can create in which no digit is repeated?

Calculate This!

Use the numbers in the grid below to answer the following questions. The same digit may not be used more than once in each number.

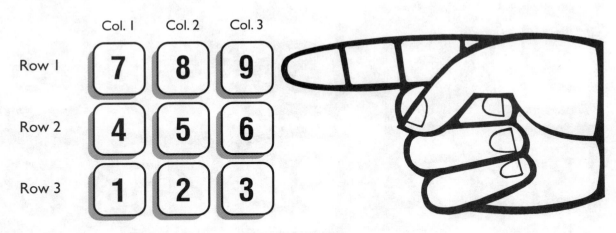

	Col. 1	Col. 2	Col. 3
Row 1	7	8	9
Row 2	4	5	6
Row 3	1	2	3

_____ 1. What is the difference between the largest three-digit number and the smallest three-digit number made from the numbers in Row 1?

_____ 2. What is the difference between the largest four-digit number made from the four corners and the smallest three-digit number made from Column 3?

_____ 3. What is the difference between the smallest three-digit number made from Column 2 and the largest two-digit number made from Row 2?

_____ 4. What is the product of the smallest three-digit number made from Row 3 and the largest two-digit number made from Column 2?

_____ 5. Without using any of the numbers in the four corners, find the sum of all three-digit numbers you can make that lie between 600 and 650.

_____ 6. Using only Column 3, what is the sum of the five smallest three-digit numbers you can make?

_____ 7. Using only the four corner keys, what is the sum of the five largest two-digit numbers you can create?

_____ 8. Using only the numbers in the two diagonals, find the sum of the five largest four-digit odd numbers you can make.

Elephant Estimation

How much do you know about elephants—specifically, African bull elephants. Fill in a number from the right-hand column to complete the statements on the left. Use reference books or the Internet for help.

1. An angry elephant can charge at _____ miles per hour.

2. A large bull elephant may be _____ feet tall at the shoulder.

3. A grazing elephant may eat _____ pounds of food in a day.

4. The trunk of an elephant holds _____ quarts of water.

5. At birth, an elephant is _____ inches tall.

6. An elephant's ear is six feet wide and weighs _____ pounds.

7. The gestation period for elephants is _____ months.

8. There are _____ muscles and tendons in an elephant's trunk.

9. A female elephant usually produces one calf every _____ years.

10. An elephant's trunk is about _____ feet long.

11. A tusk that has grown to a record length of twelve feet weighs _____ pounds.

12. An elephant may live for _____ years, or even longer.

13. The tail of an elephant is _____ feet long.

14. Elephants walk about _____ miles per hour.

15. An African bull elephant may weigh _____ pounds.

3
4
5
5
8
12
22
25
36
60
110
200
350
15,000
40,000

Calendar Capers

Use what you know about a calendar, without peeking at one, to help answer the following questions:

_____ 1. During what month does the 100th day of the year fall?

_____ 2. During what month does the 200th day of the year fall?

_____ 3. During what month does the 300th day of the year fall?

_____ 4. Name the only two consecutive months during a calendar year that have the same number of days.

_____ 5. Name three consecutive months that have a total of 91 days.

_____ 6. What is the greatest total number of days in three consecutive months?

_____ 7. How many more months have 31 days than have 30?

_____ 8. During a leap year, how many more days do the last six months of the year have than the first six?

_____ 9. What is the total number of days in the months whose names have three syllables?

_____ 10. What is the total number of days in the months whose names begin with a vowel?

Shady Deal

Study the completed Venn Diagram below. Use the numbers on the shapes on the right side of the page to fill in the remaining five diagrams. Place the numbers 1 through 12 in the correct section of each diagram.

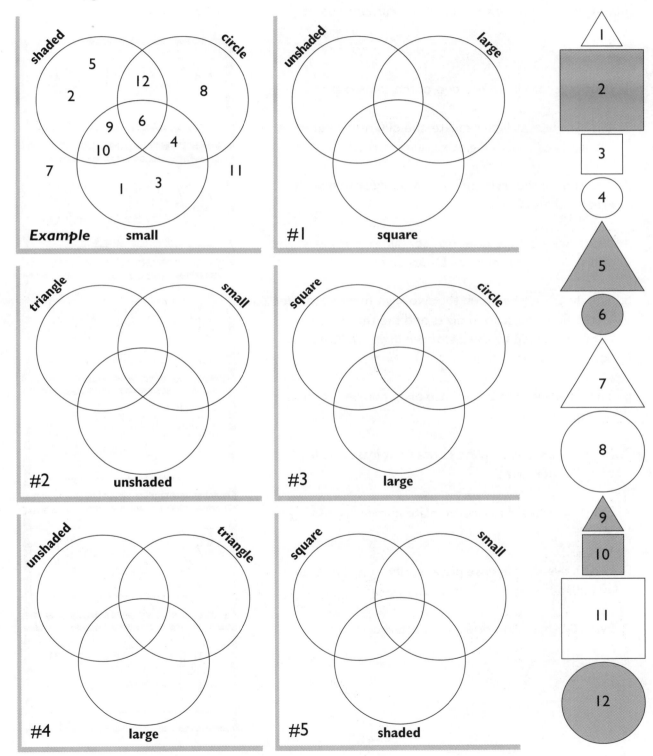

License Plate Mix-Up

The annual Middletown Road Race is scheduled for next Saturday. Each racer has been issued an identification tag in the form of a special license plate that must be displayed in the rear window of the car during the race. Your job is to match up each license plate with its owner using the clues below. Write each racer's name on the correct plate.

1. Leigh's license plate has two of the same digit.

2. On Marchessa's license plate, the difference between the first and fourth digits is greater than 1.

3. The sum of the digits on Donovan's license plate is greater than 20.

4. Troy's license plate does not contain any of the letters in his name. Neither does Donovan's.

5. The difference between the two-digit number formed by the first and second digits and the three-digit number formed by the last three digits on Kristin's license plate is even.

6. The last digit of Evan's license plate number is not prime.

7. Marchessa's license plate contains at least one letter that is in her name.

8. Donovan's five-digit license plate number is less than 50,000.

9. Leigh's five-digit license plate number is more than 50,000, but Kristin's is not.

10. Evan's license plate does not contain a zero.

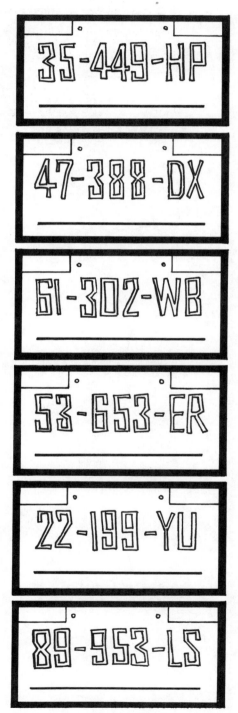

That's Odd

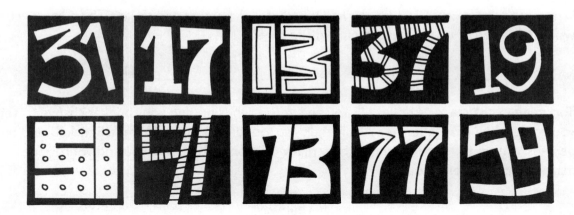

Use the two-digit numbers shown above to help answer the following questions:

_____

_____

_____

_____

_____

_____

1. What is the sum of the prime numbers shown above?

2. Which three numbers added together give a sum closest to 150 without going over?

3. Which two numbers multiplied together give a product closest to 2,500 without going over?

4. What is the sum of all the numbers whose digits are both prime?

5. Use pairs of the two-digit numbers above to create four-digit numbers. What is the sum of the three largest four-digit numbers you can create without using any two-digit number more than once?

6. What is the sum of *all* the odd four-digit numbers you can create using pairs of the two-digit numbers above—if no digit is repeated within any four-digit number?

Money Problems

1. Lizette was saving coins. In a jar, she had **98** coins—all pennies and nickels. The coins were worth $2.78. How many of each coin did she have?

_____ pennies, _____ nickels

2. Tori needed change for a dollar bill. You and your friends pooled your pocket change to see if you could make change for her. Altogether, you found 3 quarters, 6 dimes, 8 nickels, and 9 pennies. Without using the pennies, how many different ways could you and your friends make change for Tori's dollar?

_____ different ways

3. Mrs. Garner offered her students four options for earning money and asked each student to choose one. Here's what they chose:

Marla: 1¢ the first day; each day after that, the previous day's earnings are to be doubled.

Glen: 1¢ the first day, 10¢ the second day, 1¢ the third day, 10¢ the fourth day; and so on.

Sallie: a penny the first day, a nickel the second day, a dime the third day, a penny the fourth day, a nickel the fifth day; and so on.

A. After 5 days, who had the most money? _____

B. After 10 days, who had the most money? _____

C. What was the difference between the highest and lowest totals after 10 days? _____

4. Five brothers each have some coins. Among the boys, they have a dollar's worth of quarters, but no pennies. Each boy has exactly sixty cents. Jimmy has twice as many coins as Stephen, and J.R. has twice as many coins as Jimmy. Tim has five coins, which is three fewer than Josh. Which coins does each boy have?

Jimmy: _____

Josh: _____

J.R. _____

Stephen: _____

Tim: _____

Keyboard Math

Use the three rows of letters from a standard computer keyboard to help you brainstorm words for the following categories. Each letter has the point value shown.

Row 1 | Q W E R T Y U I O P | 1 point each

Row 2 | A S D F G H J K L | 2 points each

Row 3 | Z X C V B N M | 3 points each

Shaded keys are typed with the left hand; unshaded keys with the right.

Find a word that…

RULE	WORD	SCORE
1. is typed with the left-hand keys only and worth more than 8 points.	_____	_____
2. is a 5-letter word whose value is a prime number.	_____	_____
3. is worth more than 8 points and uses only keys in row 2.	_____	_____
4. is worth less than 5 points and is typed with both hands.	_____	_____
5. has 3 syllables and is worth 10–15 points.	_____	_____
6. is a proper noun typed with only right-hand keys and worth 5–10 points.	_____	_____
7. is a 3-letter word worth 6–9 points.	_____	_____
8. has 6 letters and uses only letters from rows 1 and 3.	_____	_____
9. has 2 syllables and is typed with only right-hand keys.	_____	_____
10. has 2 syllables and is typed with only left-hand keys.	_____	_____

Three-Digit Tricks

Find the largest three-digit number…

1. that is an *even* palindrome. _____

2. that is a multiple of 8. _____

3. whose first two digits add up to the third. _____

4. that isn't evenly divisible by 2, 3, or 4. _____

5. whose digits are all prime. _____

6. whose digits are all even. _____

7. whose digits have a sum of 16. _____

8. that is divisible by both 3 and 10. _____

9. that is a multiple of 27. _____

10. whose last two digits add up to the first. _____

11. that is a square number. _____

12. whose digits have a sum of 23. _____

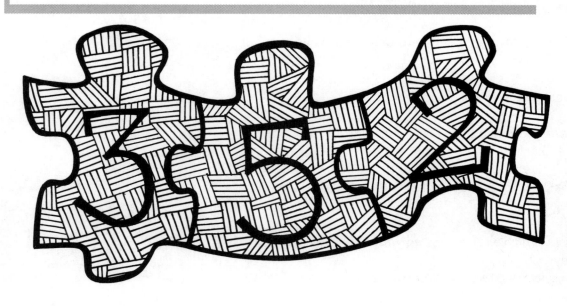

Factor This

_____ 1. What is the sum of all of the odd factors of 25?

_____ 2. What is the sum of all of the odd factors of 70?

_____ 3. What is the sum of all of the factors of 30?

_____ 4. What is the sum of all of the factors of 48?

_____ 5. What is the sum of all of the even factors of 60?

_____ 6. What is the sum of all of the even factors of 88?

_____ 7. What is the product of all of the factors of 10?

_____ 8. What is the product of all of the factors of 18?

_____ 9. What is the product of all of the factors of 40?

_____ 10. What is the sum of all of the two-digit factors of 100?

_____ 11. What is the difference between the largest *even* two-digit factor and the largest *odd* two-digit factor of 1000?

_____ 12. What is the difference between the GCF and the LCM of 24 and 120?

_____ 13. What is the difference between the GCF and the LCM of 55 and 121?

_____ 14. What is the difference between the GCF and the LCM of 18 and 360?

Day After Day

Solve the following brainteasers.

1. If yesterday was Wednesday, what will be the day after the day after tomorrow?

   _____

2. If three days from now is Thursday, what was the day before the day before yesterday?

   _____

3. Yesterday was the 13th. If the day after the day after tomorrow is Wednesday, what will the date be on Saturday?

   _____

4. If four days from now will be Monday, what was the day after the day before yesterday?

   _____

5. My birthday was five days ago. If three days after the day after tomorrow is the 27th, what was the date of my birthday?

   _____

6. Yesterday was Sunday. What was five days before the day after the day after tomorrow?

   _____

7. Five days from now will be Friday. What day was four days before the day after tomorrow?

   _____

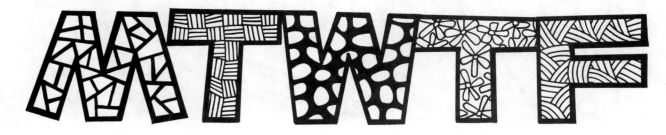

House Call

Anna, Barb, Carl, Dale, Erin, and Fred all live on the same street. Use the clues below to figure out who lives in which house.

Clues:

- Barb lives between Fred and Dale.

- The sums of the digits in Erin's and Anna's house numbers are the same.

- The digits in Fred's house number are all prime.

- The digits of Dale's and Carl's house numbers are in ascending order.

- Dale lives next door to Barb, but not to Carl.

- One of the houses next to Anna's is for sale.

Solution:

Anna lives at number _____   Dale lives at number _____

Barb lives at number _____   Erin lives at number _____

Carl lives at number _____   Fred lives at number _____

Digital Dilemma

On a digital clock, the hour is on the left and the minutes are on the right. They are separated by a colon. Use a digital clock display like the one below to help answer the following questions:

1. List five times whose digits have a sum of 10. (Example: 3:43)

   _____

2. List five times at which the sum of the minutes digits equals the hour. (Example: 10:55)

   _____

3. List five times at which the product of the minutes digits equals the hour. (Example: 4:22)

   _____

4. Some times, such as 2:02, are palindromes (the same forward and backward). List all palindromic times whose digits add up to more than 18. (Example: 8:58)

   _____

5. List all times whose digits add up to 3.

   _____

6. List all times whose digits are all prime numbers less than 5.

   _____

Go with the Flow

Start with the input number. Use the key below to help you follow it through the chart until you arrive at an output number. Write the output number in the chart below.

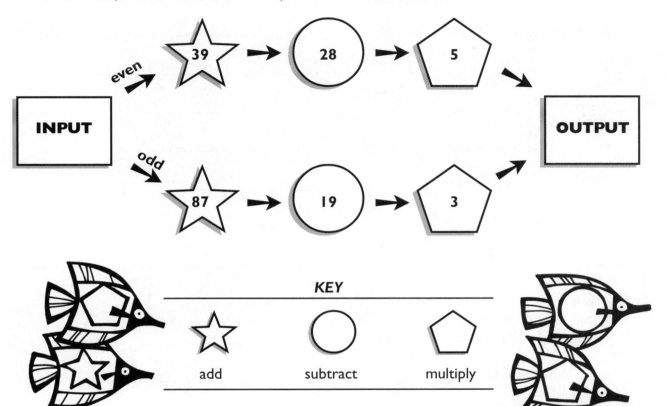

	Input	Output		Input	Output
1.	29	_____	6.	126	_____
2.	36	_____	7.	150	_____
3.	88	_____	8.	203	_____
4.	73	_____	9.	235	_____
5.	94	_____	10.	562	_____

Time Flies

What time is it when it is...

1. 37 minutes before half past 5:00 P.M.?   _____

2. 93 minutes past 4:15 A.M.?   _____

3. 211 minutes before 2:45 P.M.?   _____

4. 14$\frac{1}{2}$ hours before 10:32 A.M.?   _____

5. 404 minutes past 4:04 P.M.?   _____

6. 111 minutes past 11:11 A.M.?   _____

7. 75 hours before 7:05 P.M.?   _____

8. 700 minutes past 7:00 P.M. Wednesday?   _____

9. 12 hours and 12 minutes past 12:12 P.M.?   _____

10. 133 minutes past 1:33 A.M.?   _____

11. 606 minutes before 6:06 A.M.?   _____

12. 11 hours and 11 minutes before 11:11 A.M.?   _____

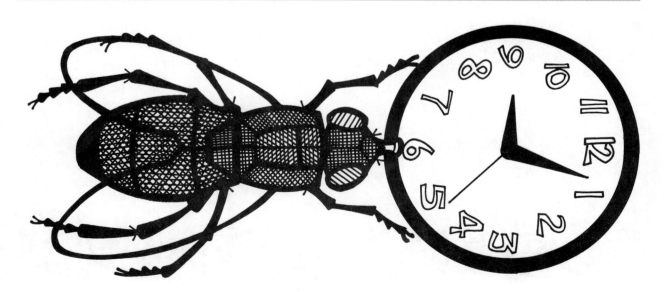

Stack 'Em Up

Use the clues below to determine where each family lives in the apartment building.
Write each family's name on the correct line.

Clues:

- The Chens live somewhere between the Thorpe and Ramirez families.
- The Vance family lives on an even-numbered floor.
- The Masters family lives three floors below the Ramirez family.
- Neither the Nortons nor the Hansons live on the ground floor.
- Only two families live above the Thorpes.
- The Smiths have to climb four flights of stairs when the elevator isn't working.
- Marla Hanson lives above her best friend, Kylie Thorpe.
- The Lucas family lives on a higher floor than the Smiths.
- The Vance family lives four floors below the Thorpe family.
- The Chen family lives on a floor that is a prime number.
- The Ramirez family does not live on the top floor.
- The Norton family lives five floors below the Chens.
- The Masters family lives on a higher floor than the Arnaz family.

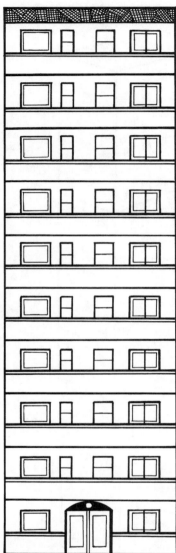

Solution:

10th floor: _____   5th floor: _____

9th floor: _____   4th floor: _____

8th floor: _____   3rd floor: _____

7th floor: _____   2nd floor: _____

6th floor: _____   1st floor: _____

Name _____

No Dice!

Observe the regular six-sided dice stacked on the left. Use what you can see to help you figure out what you can't see. Remember that although each is turned to expose different sides, all the dice are identical.

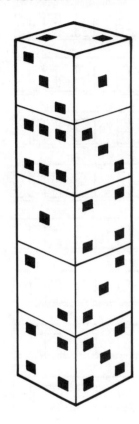

_____ 1. What is the sum of the dots on the tops of the dice?

_____ 2. What is the sum of the dots on the bottoms of the dice?

_____ 3. What is the sum of the dots on the backs of the dice?

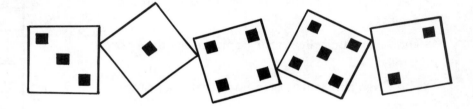

Use what you know about regular dice to answer the following questions.

_____ 1. When you roll two dice and get 8, what will always be the total of the bottoms of the dice?

_____ 2. When you roll two dice and get 5, what will always be the total of the bottoms of the dice?

_____ 3. If three dice were stacked so that each had 1 on the front, what would be the highest possible total of the tops of the three dice?

Follow the Leader

1. Find 5 consecutive numbers that add up to 100.

   _____

2. Find 7 consecutive numbers that add up to 777.

   _____

3. Find 3 consecutive even numbers that add up to 3,000.

   _____

4. Find the 3 consecutive prime numbers that add up to 97.

   _____

5. Find 3 consecutive multiples of 9 that add up to 243.

   _____

6. Find 5 consecutive numbers that add up to 555.

   _____

7. Find 3 consecutive even numbers that add up to 666.

   _____

8. Find the 4 consecutive prime numbers whose sum comes as close as possible to 100 without going over.

   _____

Bingo!

Use this Bingo card to help you answer the questions below.

	B	I	N	G	O	TOTAL
Row 1	7	18	32	59	70	_____
Row 2	15	16	31	50	62	_____
Row 3	4	24	FREE	49	74	_____
Row 4	9	30	42	47	68	_____
Row 5	8	27	39	60	66	_____
TOTAL	_____	_____	_____	_____	_____	

1. Give the total for each row and column on the lines provided above.

2. What is the difference between the totals of the highest and lowest rows?

_____

3. What is the difference between the column with the highest total and the row with the lowest total?

_____

4. What number would need to be removed from the I column to make the total 99?

_____

5. How much more does the G column total than Row 4?

_____

Name _____

Four-Digit Fun

Find the largest four-digit number…

1. that is a multiple of 7. _____

2. that has four different odd digits. _____

3. that is a palindrome. _____

4. that has no odd digits. _____

5. that is evenly divisible by 12. _____

6. whose digits have a sum of 18. _____

7. whose digits are all different. _____

8. that is a multiple of 32. _____

9. that is evenly divisible by 2, 4, and 5. _____

10. that is a square number. _____

11. whose digits are all prime. _____

12. that has both 25 and 30 as factors. _____

Counting Cookies

All of the cookies in the rectangle contain marshmallows. All of the cookies in the circle contain walnuts. All of the cookies in the triangle contain coconut. Use this information to answer the following questions:

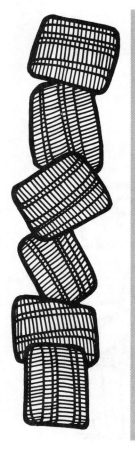

How many cookies contain...

_____ 1. only marshmallows?

_____ 2. only walnuts?

_____ 3. only coconut?

_____ 4. only walnuts and marshmallows?

_____ 5. only coconut and walnuts?

_____ 6. coconut, marshmallows and walnuts?

_____ 7. only marshmallows and coconut?

_____ 8. no walnuts?

_____ 9. no coconut?

_____ 10. no marshmallows?

Secret Code

Each of the numerals 0 through 9 is represented by a different letter of the alphabet. Use the clues to help determine which numeral each digit stands for.

Clues:

$t + t = c$
$n - m = n$
$z + t = xm$
$p + c = xx$
$d - h = h$
$h + h + h = w$

Solution:

0 = _____

1 = _____

2 = _____

3 = _____

4 = _____

5 = _____

6 = _____

7 = _____

8 = _____

9 = _____

Complete with the appropriate letters:

1. $xx - d =$ _____

2. $z \times c =$ _____

3. $z + z + z =$ _____

4. $n + p + x =$ _____

5. $z - d =$ _____

6. $w + h =$ _____

7. $t + z + n =$ _____

8. $nn - xn =$ _____

9. $n \times m =$ _____

10. $h + c =$ _____

27

A-"maze"-ing Pathways

Begin with the "start" number. Move vertically or horizontally through the maze, one square at a time. Use repeated addition or subtraction to reach the "finish" number at the end of the path. Draw a line to show your path. No square may be used more than once, and some squares won't be used at all. You must end with the total in the "finish" box.

START 1	11	25	12	36	
35	10	37	4		
24	11	3	8		
14	22	13	3	100 FINISH	

START 100	8	16	9	14	
12	27	32	11		
17	3	6	26		
4	13	7	17	1 FINISH	

START 8	23	18	42	16	
35	19	14	62		
29	31	27	58		
40	9	54	36	222 FINISH	

Pick a Digit

Use the numerals shown below, no more than once each per answer, to answer the following questions:

_____ 1. What is the difference between the largest and smallest 4-digit numbers you can make?

_____ 2. What is the sum of the three largest 3-digit numbers you can make?

_____ 3. What is the difference between the largest odd 3-digit number and the smallest odd 3-digit number you can make?

_____ 4. What is the sum of the 5 largest 2-digit numbers you can make?

_____ 5. What is the product of the largest 5-digit number and the smallest 2-digit number you can make?

_____ 6. What is the sum of the 3 largest odd 3-digit numbers you can make?

_____ 7. What is the difference between the largest even 5-digit number and the smallest odd 5-digit number you can make?

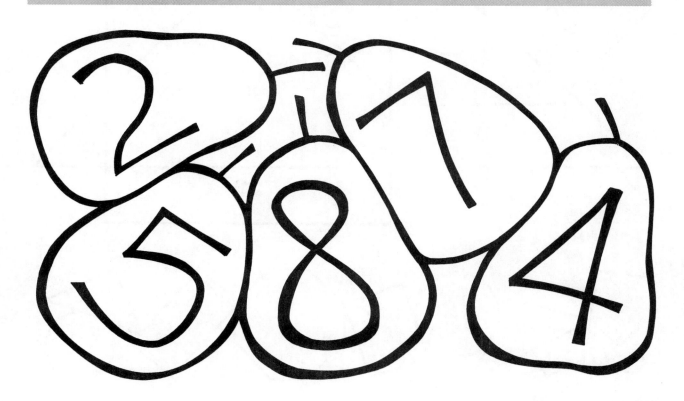

Shape Stumpers

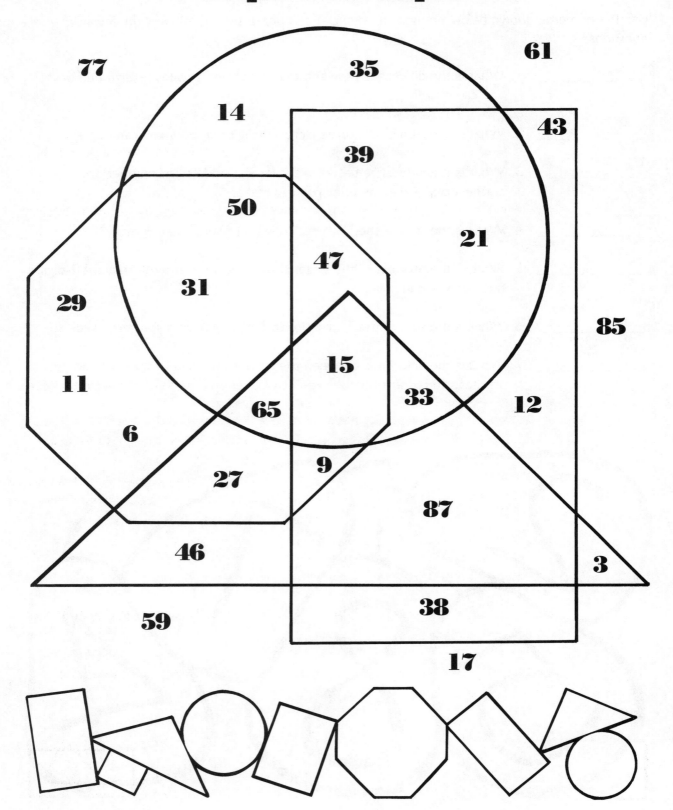

Shape Stumpers

Use the numbers that lie inside and outside the overlapping shapes—rectangle, circle, triangle, and octagon—to answer the following questions:

_____ 1. Which number lies in all four shapes at the same time?

_____ 2. How many numbers lie in only one shape?

_____ 3. What is the difference between the sum of the numbers that lie outside all the shapes and the sum of the numbers that lie within the octagon?

_____ 4. Which shape has the smallest sum within its borders—the rectangle, circle, triangle, or octagon?

_____ 5. What is the product of the two largest numbers that lie outside all the shapes?

_____ 6. What is the sum of all the two-digit numbers that lie anywhere inside any of the four shapes?

_____ 7. What is the sum of the numbers that lie within the circle, but *not* within the rectangle?

_____ 8. What is the sum of all the numbers that lie anywhere outside the triangle?

_____ 9. What is the difference between the sum of the two-digit *even* numbers that lie anywhere inside the shapes and the sum of the two-digit *odd* numbers that lie anywhere inside the shapes?

_____ 10. What is the difference between the sum of the numbers in the triangle and the sum of the numbers in the circle?

_____ 11. What is the sum of the numbers that lie in exactly two of the shapes at the same time?

_____ 12. What is the product of the largest even number in the triangle and the largest odd number in the circle?

MATH CRITICAL THINKING
Copyright © 2004 The Learning Works

Map Mysteries

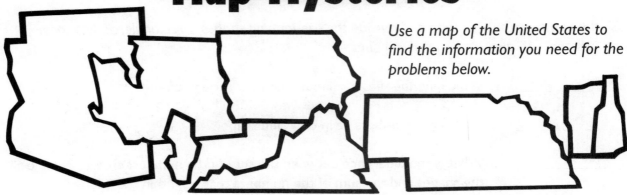

Use a map of the United States to find the information you need for the problems below.

_____ 1. The number of letters in the name of California's state capital, times the number of letters in the name of the state directly north of Oregon, times the number of states that touch the Pacific Ocean, times the number of states that touch Canada.

_____ 2. The number of letters in the state directly south of New Hampshire, times the number of letters in the capital city of the state directly south of Georgia, times the number of letters in the name of the state that separates Kentucky from Alabama.

_____ 3. The number of letters in the name of the state directly north of Kansas, times the number of letters in the name of the state directly south of Arkansas, plus the number of letters in the name of Maryland's capital, minus the number of letters in the name of the state whose capital is Boise.

_____ 4. The number of letters in the name of the capital of Indiana, times the number of letters in the name of Oregon's capital, plus the number of letters in the state that is directly east of Kansas, plus the number of letters in the state directly east of North Dakota.

_____ 5. The number of states that touch the Great Lakes, times the number of states that touch the Gulf of Mexico, times the number of states that touch Mexico, times the number of states that touch Oklahoma.

_____ 6. The number of states with two-word names, times the number of state capitals with two-word names, times the number of states whose names begin with a vowel, plus the number of states whose names end with a vowel, plus the number of states with two-syllable names. (Do not count "y" as a vowel.)

Heads Up!

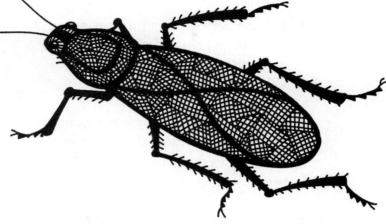

_____ 1. Tara worked at the pet store. One morning several of the cage doors had been left open. The puppies and canaries were loose. Animals were everywhere! She spotted 30 heads in the crowd, and 84 feet. How many of the escapees were canaries?

_____ 2. At the fix-it shop, Mike was surrounded by bicycles and tricycles that had been brought in for repair. He counted 24 seats and 54 wheels. How many tricycles had been brought in?

_____ 3. The tarantulas and cockroaches from Jamie's critter collection had gotten loose. From her vantage point on top of the kitchen table, Jamie's mom spotted 15 heads and 102 legs. How many of the critters were cockroaches?

_____ 4. At the junkyard, Jarrett was looking for parts to build a racer. He was looking in the area where the old cars and bicycles were parked. Rummaging around, he found 19 vehicles. They had 60 tires, most of which were flat. How many of the vehicles were cars?

_____ 5. This month, a special exhibit at the zoo featured black-and-yellow animals. There were bumblebees and goldfinches in one display. In all, there were 25 animals. Among them, they had 78 legs. How many goldfinches were in the exhibit?

_____ 6. Some male narwhals and walruses were frolicking in the sun on and around a giant iceberg. Observers counted 60 heads and 100 tusks. How many of the animals were walruses?

Measure Up

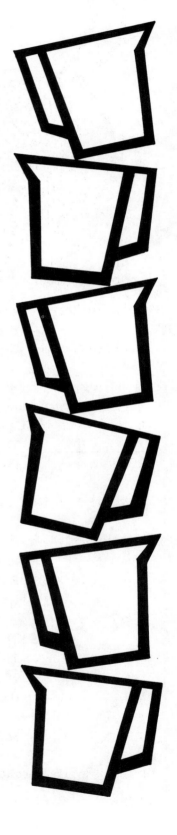

_____ 1. The number of feet in three miles minus the number of inches in twelve yards, plus the number of rods in six furlongs.

_____ 2. The number of pints in twenty quarts, plus the number of pecks in fourteen bushels, plus the number of cups in two gallons.

_____ 3. The number of square feet in two square yards, plus the number of square inches in five square feet.

_____ 4. The number of minutes in three days, plus the number of hours in five weeks, plus the number of days in two years.

_____ 5. The number of weeks in six years, plus the number of years in fourteen decades, minus the number of decades in six centuries.

_____ 6. The number of centimeters in six meters, plus the number of meters in six kilometers, plus the number of milligrams in six grams.

_____ 7. The number of yards in a half-mile, plus the number of inches in three-and-a-half yards, plus the number of feet in three-quarters mile.

_____ 8. The number of quarters in eighteen dollars, plus the number of nickels in fourteen quarters, plus the number of dimes in sixty-three dollars.

Five-Digit Fun

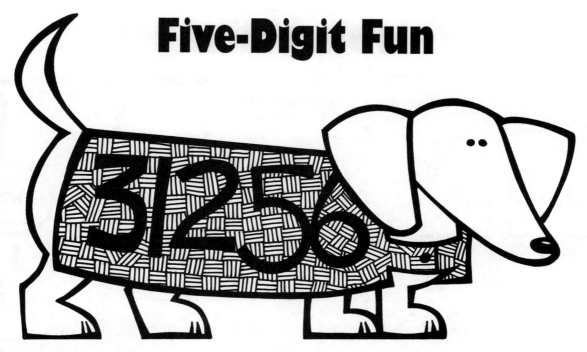

What is the largest five-digit number...

1. whose digits are all prime? _____

2. that is a palindrome? _____

3. not divisible by 2, 3, 4, or 5? _____

4. whose first digit is the sum of the last four? _____

5. whose digits have a sum of 23? _____

6. that is a multiple of 75? _____

7. that is divisible by both 24 and 50? _____

8. whose digits are all different? _____

9. whose last digit is the sum of the first four? _____

10. that is a square number? _____

11. whose five different digits have a sum of 17? _____

12. that is even and contains four different prime digits? _____

Flowing Along

Place each input number in the flow chart. Follow the arrows and complete each operation. When you arrive at an output number, place it in the chart below.

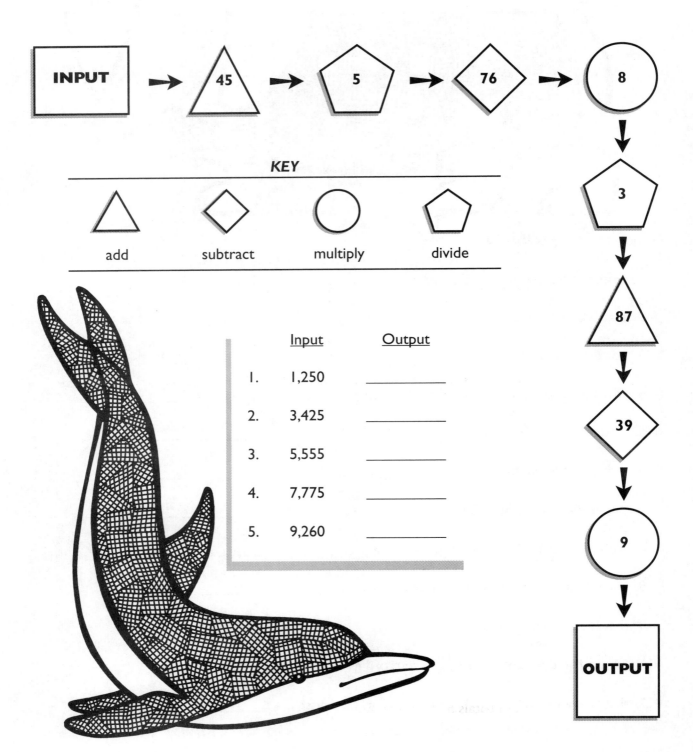

INPUT → 45 → 5 → 76 → 8

KEY

△	◇	○	⬠
add	subtract	multiply	divide

3

87

39

9

OUTPUT

	Input	Output
1.	1,250	_____
2.	3,425	_____
3.	5,555	_____
4.	7,775	_____
5.	9,260	_____

Name _____

Fill 'Er Up

Fill in the Bingo card using the clues below, then fill in the totals for each row and column on the lines provided.

	B	I	N	G	O	TOTAL
Row 1	6		41		68	_____
Row 2	14	17	34	48		_____
Row 3		22	FREE		74	_____
Row 4	12	30	44	57		_____
Row 5	8		38	46	63	_____

TOTAL   _____ _____ _____ _____ _____

1. The total of Row 2 is 18 more than Column N.

2. Row 4 totals 62 more than the diagonal containing the number 68.

3. Column B totals 297 less than Column O.

4. Row 3 totals 3 more than the diagonal containing the number 63.

5. Column G and Column O total 598.

6. Row 1 totals 16 less than Row 4.

7. Column I totals 53 less than Row 2.

Make a Date

The year is 2005, and May 1st falls on a Sunday. Use this information to figure out the date of each of the following events:

_____ 1. Justin Johnson's birthday falls on the third Wednesday of May.

_____ 2. Exactly three weeks before his birthday, Justin went on a Boy Scout camping trip.

_____ 3. Justin's sister has a dentist appointment sixteen days after Justin's birthday.

_____ 4. The second Saturday in June is the beginning of summer vacation.

_____ 5. Fifteen days before summer vacation begins, the school will hold its annual picnic.

_____ 6. Mr. Johnson's birthday is exactly eleven weeks before the picnic.

_____ 7. Eleven days before Mr. Johnson's birthday, he and Mrs. Johnson will celebrate their wedding anniversary.

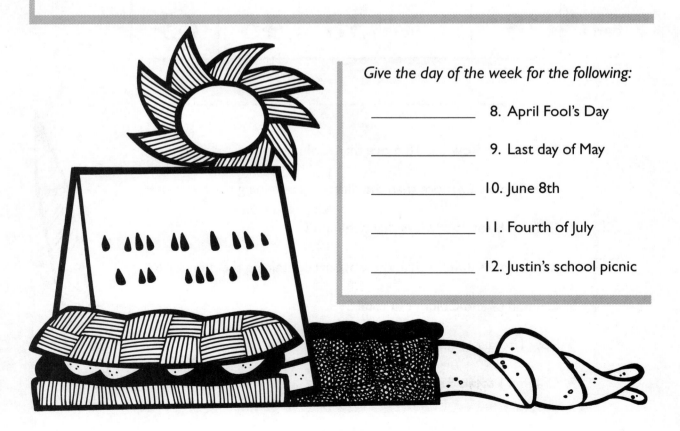

Give the day of the week for the following:

_____ 8. April Fool's Day

_____ 9. Last day of May

_____ 10. June 8th

_____ 11. Fourth of July

_____ 12. Justin's school picnic

Something in Common

What do the numbers in each set have in common? Think of another number that would fit the set, and write it in the blank at the end of the row. Then, on the line below, explain why your number fits.

Example: 752, 541, 660, 862, 945, _____

Why? The difference between the first two digits equals the third.

1. 73, 40, 84, 62, 95, _____

   Why? _____

2. 121, 33, 202, 7117, 6556, _____

   Why? _____

3. 12, 67, 345, 78, 234, _____

   Why? _____

4. 44, 53, 17, 80, 332, _____

   Why? _____

5. 24, 13, 79, 35, 57, _____

   Why? _____

6. 5, 17, 23, 31, 11, _____

   Why? _____

7. 224, 123, 325, 459, 606, _____

   Why? _____

Almanac Math

_____ 1. The year Alaska became a state, plus the year Ohio became a state, minus the number of counties in New York, minus the number of counties in Arkansas.

_____ 2. The number of stripes on Cuba's flag, times the number of stars on the flag of the Solomon Islands, times the number of stripes on the flag of Greece, times the number of colors on the flag of Sweden.

_____ 3. The year James Madison was born, plus the year Andrew Jackson took office as president, plus the year Ulysses S. Grant died, plus the year Abraham Lincoln was assassinated.

_____ 4. The height of the Washington Monument (to the nearest foot), plus the year the Lincoln Memorial was dedicated, plus the year the White House was burned by the British, minus the year the U.S. Holocaust Museum opened.

_____ 5. The year Notre Dame won its first National College Football Championship, plus the year Ohio State won its first Rose Bowl, plus the year Barry Sanders was the NFL Rookie of the Year, minus the year Texas A&M won its first Cotton Bowl.

_____ 6. The year the Winter Olympics was first held in the United States, plus the year the Summer Olympics was first held in the United States, plus the year Jesse Owens won the gold medal in the 200-meter run, minus the year Wilma Rudolph won the gold medal in the 100-meter run.

_____ 7. The year Brewster invented the kaleidoscope, plus the year Daimler invented the motorcycle, plus the year Jarvik invented the artificial heart, minus the year General Motors invented the airbag.

Name _____

Three-sided Puzzlers

Look at the first three triangles in each row to figure out the "rule" for putting numbers in the middle. Fill in the middle of the fourth triangle using the rule you discovered.

#1

2	7	3	4
4	6	8	___
4 2	8 5	6 5	9 5

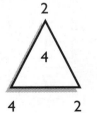   

#2

2	4	9	3
5	4	4	___
5 2	8 2	6 6	4 6

#3

4	7	4	6
8	8	7	___
8 2	6 6	9 3	8 4

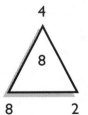 

#4

6	9	8	5
5	7	6	___
5 4	5 3	6 4	1 5

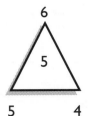  

#5

4	9	5	3
24	13	16	___
8 8	2 5	5 9	7 8

Keypad Challenge

The numbers portion of a telephone keypad is shown on the right. Each letter of the alphabet is assigned a value equal to the number on its key. Use these values to think of words that obey the following rules:

1	2 ABC	3 DEF
4 GHI	5 JKL	6 MNO
7 PQRS	8 TUV	9 WXYZ

RULE	WORD	VALUE
1. 5-letter word using only even-numbered keys	_____	_____
2. 6-letter word worth less than 20 points	_____	_____
3. 5-letter word using only the center column	_____	_____
4. 4-letter word using only odd-numbered keys	_____	_____
5. Two rhyming words using only the second row of keys	_____	_____
	_____	_____
6. 6-letter word using only prime-numbered keys	_____	_____
7. 5-letter word worth more than 25 points	_____	_____
8. 4-letter word worth more than 20 points	_____	_____
9. 4-letter word using only one key	_____	_____
10. 5-letter word worth less than 15 points	_____	_____

Alphabet Alley

At Alphabet Alley, each letter of the alphabet can be purchased for the amount shown. Use this information to find five-letter words that follow the rules below:

PRICES

A	1¢
B	2¢
C	3¢
D	4¢
E	5¢
F	6¢
G	7¢
H	8¢
I	9¢
J	10¢
K	11¢
L	12¢
M	13¢
N	14¢
O	15¢
P	16¢
Q	17¢
R	18¢
S	19¢
T	20¢
U	21¢
V	22¢
W	23¢
X	24¢
Y	25¢
Z	26¢

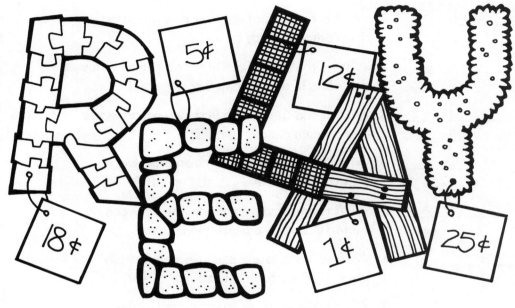

1. Costs less than a quarter _____

2. Has letters that each cost an odd number of cents _____

3. Is worth exactly a quarter _____

4. Is worth more than a dollar _____

5. Has letters whose costs are prime numbers _____

6. Has letters that are each worth less than a dime _____

7. Has letters with prices that are all multiples of three _____

8. Has letters that are each worth more than a dime _____

9. Has a total price that is a prime number _____

10. Is worth exactly half a dollar _____

Playing with Palindromes

Palindromic numbers read the same forward as backward and must have at least two digits. Examples are 22, 303, and 444. Each of the following questions involves one or more palindromes. See how many you can figure out.

_____ 1. What is the sum of the first two palindromes that contain exactly two 1's?

_____ 2. What is the sum of the first two palindromes that contain exactly two 2's?

_____ 3. What is the sum of the first two palindromes that contain exactly two 3's?

_____ 4. What is the greatest palindrome between 1 and 1000 whose digits add up to 10?

_____ 5. What is the smallest palindrome between 1 and 1000 whose digits add up to 7?

_____ 6. What is the greatest palindrome between 1 and 1000 whose digits add up to 15?

_____ 7. What is the smallest palindrome between 1 and 1000 whose digits add up to fifteen?

_____ 8. What is the largest palindrome between 1 and 1000 whose digits add up to 20?

_____ 9. What is the sum of all the palindromes between 100 and 200?

_____ 10. What is the sum of the largest and smallest three-digit palindromes that contain a zero?

_____ 11. What is the sum of the two smallest palindromes that are divisible by 3?

_____ 12. What is the sum of the two smallest palindromes that are not made up of all the same numerals?

_____ 13. What is the sum of the two smallest four-digit palindromes?

_____ 14. What is the smallest four-digit palindrome with all even numerals?

_____ 15. What is the difference between the largest and smallest four-digit palindromes?

1523768867325 1

Name _____

Probably Not!

If you were to spin the pointer on the left, what would be the probability that it would point to the following?

1. a consonant   _____

2. anything but B   _____

3. a letter with no curves   _____

If the cards below were mixed up and turned face down and one card drawn, what would be the probability of drawing the following?

1. a vowel   _____

2. a letter in the word "cabbage"   _____

3. a letter between "A" and "E"   _____

If you were to spin the pointer on the right, what would be the probability that it would point to the following?

1. an even number   _____

2. a two-digit number   _____

3. a prime number   _____

4. a number whose square is odd _____

5. a factor of 48   _____

Name _____

Math Mysteries

_____ 1. Jo left the park at 2:15. Nate left seventy-five minutes later. A half-hour after Nate left, Stuart also left the park. Five minutes before Stuart left, Ellie left. What time did Ellie leave the park?

_____ 2. Sheila had $11.20 and Tonya had $8.75. Liz had twice as much as Tonya, and Jared had half as much as Sheila. The four friends wanted to go to a concert together and needed $50.00 to pay the admission fee. How much more money did they need?

_____ 3. The desks in Allie's classroom are lined up in straight rows. When no one is absent, each desk in the room is filled. Jeannie is in the third row from the back and the third row from the front. She is also the second student from the right end of the row and the seventh student from the left end of the row. How many students are in Allie's classroom?

_____ 4. Kelsey sold cookies at her school's bake sale for twenty-five cents each. She sold two-thirds of her cookies and still had thirty-two cookies left. How much money did she make on the cookies that she did sell?

_____ 5. The Brewster kids saved dimes to donate to the pet food drive at the local animal shelter. They had $6.40 in all. Drew had the most money, which was exactly twice as much as Abigail had. Chaz had half as much as Abigail. Ben had the same amount as Abigail, plus a dollar. How much did each of them have?

Abigail: _____ Ben: _____ Chaz: _____ Drew: _____

Name _____

Tons of Numbers

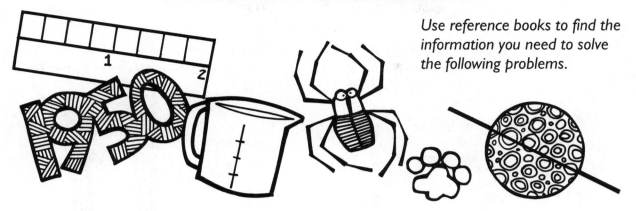

Use reference books to find the information you need to solve the following problems.

_____ 1. The number of feet in half a mile, divided by the number of legs on three spiders, times the number of sides on a pentagon, plus the number of planets in our solar system.

_____ 2. The number of inches in three yards, plus the number of seconds in ten minutes, minus the number of letters in the alphabet, plus the number of days in a leap year.

_____ 3. The number of pounds in three tons, divided by the number of days in June, minus the number of faces on a cube, times the number of cups in a quart.

_____ 4. The number of degrees in a circle, divided by the number of years in a decade, minus the number of nickels in a dollar, times the number of paws on six kittens.

_____ 5. The number of donuts in three dozen, times the number of sides on an octagon, minus the number of stripes on the flag of the United States, plus the number of singers in a barbershop quartet.

_____ 6. The number of months whose names begin with vowels, times the number of hours in a week, plus the number of legs on an octopus, minus the number of squares on a checkerboard.

Prime Time

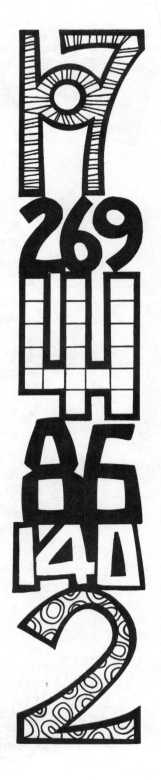

_____ 1. What is the sum of the five smallest prime numbers?

_____ 2. What is the sum of the first three two-digit prime numbers?

_____ 3. What is the product of the largest single-digit prime number and the largest two-digit prime number?

_____ 4. What is the difference between the smallest prime number and the tenth prime number?

_____ 5. What is the sum of all prime numbers between 20 and 40?

_____ 6. What is the sum of the first five prime numbers that contain a 3?

_____ 7. What is the sum of the three largest two-digit prime numbers?

_____ 8. How many prime numbers less than 100 contain an 8?

_____ 9. What is the difference between the smallest and largest two-digit prime numbers?

_____ 10. What is the product of the prime number just less than 70 and the prime number just greater than 70?

_____ 11. What is the sum of all prime numbers less than 100 that contain a 1?

_____ 12. What is the product of the two largest two-digit prime numbers that contain a 5?

Name _____

Prize Patrol

The Laketown Prize Patrol has chosen eight local citizens for their annual safe driving award. Pictures of the winners' license plates were posted at the Laketown Community Center, and winners have one week to claim their prizes. See if you can determine which winner owns each of the posted license plates using the clues below. Write name of the correct owner and the make of his or her car beneath each license plate on the following page.

WINNERS		CARS	
Scott Alderson	Steve Marsh	Dodge	Mitsubishi
Nora Bland	Harry Pendleton	Honda	Nissan
Derek Freeman	Kara Stenson	Kia	Saturn
Meg Jarvis	Lynn Talbert	Lexus	Subaru

Clues:

1. No person owns a car whose name begins with the same letter as his or her own name.

2. No license plate has any letter that is in its owner's first name.

3. Nora realized that the product of the digits on her Mitsubishi's license plate was less than their sum.

4. All of the digits on Derek's and Kara's license plates are prime numbers.

5. Both Meg and Lynn own cars with six-letter names.

6. The difference between the sum and the product of the digits on Meg's license plate is odd.

7. Scott is the only person whose car's name has fewer letters than his own name.

8. Steve and Harry are the only two whose cars' names have the same number of letters as their own names.

9. The product of the digits on Scott's license plate is greater than 100.

10. The Honda's owner realized that the difference between the sum and the product of the digits on his license plate is 40.

11. The two boys who owned the Saturn and the Lexus found that the difference between the three-digit numbers on their license plates is 269.

12. The three-digit number on the Nissan's license plate is greater than the three-digit number on the Subaru.

Prize Patrol (continued)

Make: _____

Owner: _____

Make: _____

Owner: _____

Make: _____

Owner: _____

Make: _____

Owner: _____

Make: _____

Owner: _____

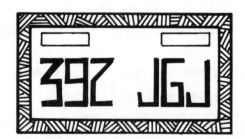

Make: _____

Owner: _____

Make: _____

Owner: _____

Make: _____

Owner: _____

Calculator Products

A calculator and trial and error will help you figure out these problems! Using the digits given for each problem, figure out which arrangement gives you the greatest and least products.

	DIGITS	LEAST PRODUCT	GREATEST PRODUCT
1.	2, 8, 6, 3	___ ___ × ___ ___ _____	___ ___ × ___ ___ _____
2.	6, 6, 8, 9	___ ___ × ___ ___ _____	___ ___ × ___ ___ _____
3.	5, 7, 2, 8	___ ___ × ___ ___ _____	___ ___ × ___ ___ _____
4.	9, 2, 6, 7	___ ___ × ___ ___ _____	___ ___ × ___ ___ _____
5.	4, 6, 7, 2, 9	___ ___ ___ × ___ ___ _____	___ ___ ___ × ___ ___ _____
6.	3, 7, 9, 2, 8	___ ___ ___ × ___ ___ _____	___ ___ ___ × ___ ___ _____

Go Figure!

Use your "math logic" to help you figure out the following riddles:

_____ 1. The winning race car had a three-digit odd number painted on its side. The sum of the digits was 12. None of the digits was the same. The hundreds digit was twice the ones digit. What was the number?

_____ 2. The number on Carl's house is a four-digit even number whose digits are all different. No digit is 0. The number is more than 1000, but less than 2000. The sum of the digits is 16. The hundreds digit is three more than the ones digit. What is the number?

_____ 3. Suzi collects books. She has an even number of books in her collection. The sum of the digits in the number of books is fifteen. The tens digit is twice the hundreds digit. If she has between 2000 and 3000 books in her collection, how many books does she have?

_____ 4. Tory was trying to guess a four-digit number that was hidden beneath four cards. She was given the following clues:
   • each digit is smaller than the one preceding it
   • the number is odd and between 8000 and 9000
   • the two-digit number formed by the thousands and hundreds digits is 33 more than the two-digit number formed by the tens and ones digits.
   • the sum of the digits is 22

Exploring the Americas

Use a world atlas or almanac to find the information you need to solve the following problems.

_____ 1. The number of states that touch Mexico, plus the number of countries between Honduras and Colombia, times the number of letters in the names of the countries touching Lake Titicaca.

_____ 2. The number of Canadian provinces touching the state of Maine, plus the number of South American countries through which the equator passes, times the number of letters in the name of the Mexican state in which Acapulco is located.

_____ 3. The number of letters in the name of the country between Chile and Uruguay, times the number of provinces between British Columbia and Ontario, plus the number of states that touch Oklahoma.

_____ 4. The number of letters in the name of Belize's capital, times the number of letters in the name of the country directly south of Florida, minus the number of U.S. states whose names begin with the letter "A."

_____ 5. The number of letters in the name of the country whose capital is Kingston, times the number of Mexican states that touch Guatemala, times the number of vowels in the name of Canada's capital.

_____ 6. The number of colors on Puerto Rico's flag, plus the altitude in feet of California's Mount Whitney, minus the number of states touching Lake Erie.

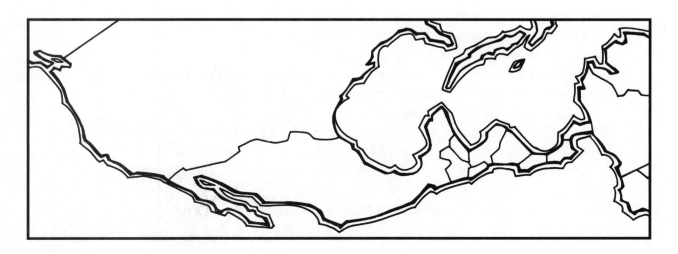

Feline Fun

The Kennedy Kat Klub recently held its annual Feline Fair. There were eight entries in the "Best Dressed Cat" category. Unfortunately, the cats' ID tags got mixed up. Use the clues below to match each entrant to his or her tag. Write each cat's name beneath the correct tag.

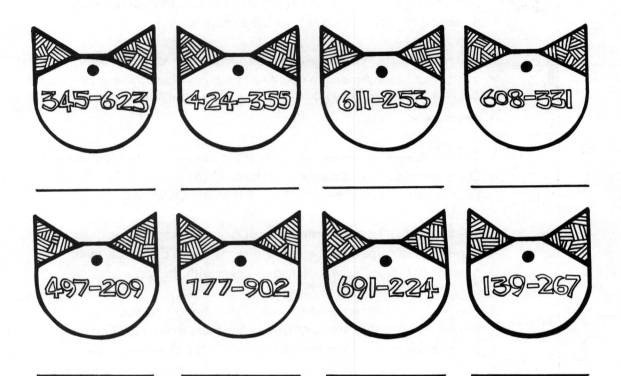

Clues

- The sum of the two three-digit numbers in Flair's ID is odd.
- The sum of the two three-digit numbers in both Fillis's and Flash's ID numbers is a three-digit number.
- Fred's ID number has at least two digits the same.
- The sum of the digits on Fuzzy's tag is prime.
- The difference between the two three-digit numbers on Finicky's tag is odd.
- The sums of the digits on Fuzzy's and on Fifi's tags are the same.
- Fluffy has at least one 3 on her tag.
- All the digits in the sum of the two three-digit numbers on Fred's tag are even.
- The cat whose ID number is 139–267 has an odd number of letters in its name.
- The difference between the two three-digit numbers on Finicky's ID number is greater than 300.
- Fifi's tag has two pairs of matching numbers.

Name _____

Don't Be Square

1.

2.

3.

4.

5.

6.

7.

8.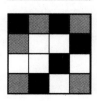

_____ 1. If the perimeter of the small square is 24 and the perimeter of the large square is 48, what is the area of the unshaded portion of the figure?

_____ 2. If the area of the shaded region is 147, what is the perimeter of the square?

_____ 3. The unshaded square has an area of 144. Each corner of the shaded square lies on the midpoint of a side of the unshaded square. What is the area of the shaded square?

_____ 4. If the diameter of the circle is 10, what is the area of the shaded portion of the figure?

_____ 5. If the area of the large square is 400, what is the sum of the areas of the small unshaded squares?

_____ 6. The area of the square is 256. The radius of the circle is 1. At most, how many circles will fit inside the square without overlapping?

_____ 7. Each small square has an area of 36. The sixteen small squares can be arranged to form rectangles of various sizes. What is the greatest perimeter that can be achieved?

_____ 8. If the small black, gray, and white squares were turned face-down, how many would you have to turn over to be sure that you had at least one black square?

55

A-B-C Math

Each letter of the alphabet can be purchased for the amount shown. Use this information to answer the following questions:

PRICES

A	1¢
B	2¢
C	3¢
D	4¢
E	5¢
F	6¢
G	7¢
H	8¢
I	9¢
J	10¢
K	11¢
L	12¢
M	13¢
N	14¢
O	15¢
P	16¢
Q	17¢
R	18¢
S	19¢
T	20¢
U	21¢
V	22¢
W	23¢
X	24¢
Y	25¢
Z	26¢

1. Write a five-word sentence that Sarah could purchase for less than a dollar.

   _____

2. What animal name could John buy that would cost more than a dollar?

   _____

3. Would it cost more to buy the name of your city or of your state? How much more?

   _____

4. Pam challenged Sharon to find a word worth exactly one dollar. What word could Sharon use?

   _____

5. Write a word with all different letters that costs exactly the same as the word "dim."

   _____

6. Which word in each category is worth closest to one dollar without going over?

   a. Great Lake: _____ value: _____ ¢

   b. continent: _____ value: _____ ¢

   c. ocean: _____ value: _____ ¢

   d. month: _____ value: _____ ¢

   e. day of the week: _____ value: _____ ¢

   f. five states with largest areas: _____ value: _____ ¢

Name _____

It's a Dog's Life

The town of Dogpatch has decided to institute a dog license tax to help support the local animal shelter. The first eight dogs to be registered received special gold tags for their collars. Match each dog with his or her tag, and write the dog's name below his or her tag.

CBT-841   YWE-118   SSH-362   KLP-136

_____   _____   _____   _____

ACT-212   XRT-567   NNO-782   LJD-407

_____   _____   _____   _____

Clues:

- The sum of the digits on Black Jack's tag is prime.
- Buzzy's tag has two matching digits.
- The three-digit number on Bart's tag is even.
- Bernice's tag has at least one letter that is in her name.
- There is a difference of 431 between Bojo's and Brownie's three-digit tag numbers.
- The sum of the digits on Beast's tag is less than 12.
- Buzzy's tag contains no prime-number digits.
- The dog whose tag reads SSH–362 has an odd number of letters in its name.
- The digits on Benji's tag are in descending order.
- Bart's tag has none of the letters in his name.
- Brownie's three-digit tag number is smaller than Bojo's.
- There is a vowel on Bernice's tag.
- Beast's tag does not contain a zero.

Answer Key

Note: Answers are given to show examples that meet question criteria and to show possible correct answers. Some activities may have additional correct answers. Give students credit for any answers that match the criteria given.

Page 5: Twice as Nice

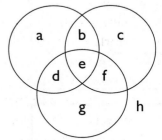

1. 41 & 56
2. 34 & 41
3. 123
4. 7,012
5. 1,857,945
6. 42
7. 392
8. $927,856 - 234,156 = 693,700$

Page 6: Calculate This!

1. 198
2. 9,362
3. 193
4. 10,455
5. 3,812
6. 3,033
7. 433
8. 48,685

Page 7: Elephant Esimation

1. 25
2. 12
3. 350
4. 5
5. 36
6. 110
7. 22
8. 40,000
9. 4
10. 8
11. 200
12. 60
13. 5
14. 3
15. 15,000

Page 8: Calendar Capers

1. April
2. July
3. October
4. July & August
5. April, May, June; or September, October, November
6. 92
7. 3
8. 2
9. 122
10. 92

Page 9: Shady Deal

1. a. 1 & 4
   b. 7 & 8
   c. 5 & 12
   d. 3
   e. 11
   f. 2
   g. 10
   h. 6 & 9

2. a. 5
   b. 9
   c. 6 & 10
   d. 7
   e. 1
   f. 3 & 4
   g. 8 & 11
   h. 2 & 12

3. a. 3 & 10
   b. none
   c. 4 & 6
   d. 2 & 11
   e. none
   f. 8 & 12
   g. 5 & 7
   h. 1 & 9

4. a. 3 & 4
   b. 1
   c. 9
   d. 8 & 11
   e. 7
   f. 5
   g. 2 & 12
   h. 6 & 10

5. a. 11
   b. 3
   c. 1 & 4
   d 2
   e. 10
   f. 6 & 9
   g. 5 & 12
   h. 7 & 8

Page 10: License Plate Mix-Up

35–449–HP: Kristin
47–388–DX: Evan
61–302–WB: Troy
53–653–ER: Leigh
22–199–YU: Donovan
89–953–LS: Marchessa

Page 11: That's Odd

1. 249
2. 73 + 59 + 17 = 149
3. 77 × 31 = 2,387
4. 187 (37 + 73 + 77)
5. 21,673
6. 112,918

Page 12: Money Problems

1. 53 pennies, 45 nickels
2. 12 different ways
3. (a) Marla; (b) Marla; (c) $9.74
4. Jimmy: 6 dimes or 1 quarter, 2 dimes, and
   3 nickels; Josh: 1 quarter, 7 nickels or
   4 nickels and 4 dimes; J.R.: 12 nickels;
   Stephen: 2 quarters, 1 dime;
   Tim: 1 quarter, 3 dimes, 1 nickel

Page 13: Keyboard Math

Examples:

1. treads, dearest
2. rules, point
3. flask, glass
4. Troy, two
5. gorilla, algebra
6. Polk, Ohio
7. man, box
8. zipper, mentor
9. holly, milky
10. beaver, faster

Page 14: Three-Digit Tricks

1. 898
2. 992
3. 909
4. 997
5. 777
6. 888
7. 970
8. 990
9. 999
10. 981
11. 961
12. 995

Page 15: Factor This

1. 31
2. 48
3. 72
4. 124
5. 144
6. 168
7. 100
8. 5,832
9. 2,560,000
10. 105
11. 25
12. 96
13. 594
14. 342

Page 16: Day After Day

1. Sunday
2. Friday
3. 20th
4. Wednesday
5. 17th
6. Saturday
7. Friday

Page 17: House Call

1. Anna = 373
2. Barb = 361
3. Carl = 379
4. Dale = 367
5. Erin = 355
6. Fred = 357

Page 18: Digital Dilemma

Examples:

1. 3:34, 2:53, 4:24, 6:13, 8:02
2. 8:44, 3:12, 2:20, 7:34, 11:56
3. 3:31, 4:14, 8:24, 10:25, 7:17
4. 7:57, 8:38, 8:48, 9:19, 9:29, 9:39, 9:49, 9:59
   (must have all)
5. 1:02, 1:11, 1:20, 2:01, 2:10, 3:00, 10:11, 10:20,
   10:02, 11:10, 11:01 (must have all)
6. 2:22, 2:23, 2:32, 3:23, 3:32, 3:33 (must have all)

Page 19: Go with the Flow

1. 291
2. 235
3. 495
4. 423
5. 525
6. 685
7. 805
8. 813
9. 909
10. 2,865

Page 20: Time Flies

1. 4:53 P.M.
2. 5:48 A.M.
3. 11:14 A.M.
4. 8:02 P.M.
5. 10:48 P.M.
6. 1:02 P.M.
7. 4:05 P.M.
8. 6:40 A.M. Thursday
9. 12:24 A.M.
10. 3:46 A.M.
11. 8:00 P.M.
12. midnight

Page 21: Stack 'Em Up

10th floor: Lucas
9th floor: Hanson
8th floor: Thorpe
7th floor: Chen
6th floor: Ramirez
5th floor: Smith
4th floor: Vance
3rd floor: Masters
2nd floor: Norton
1st floor: Arnaz

Page 22: No Dice!

1. 13
2. 22
3. 19

1. 6
2. 9
3. 15

Page 23: Follow the Leader

1. 18, 19, 20, 21, 22
2. 108, 109, 110, 111, 112, 113, 114
3. 998, 1,000, 1,002
4. 29, 31, 37
5. 72, 81, 90
6. 109, 110, 111, 112, 113
7. 220, 222, 224
8. 17, 19, 23, 29

Page 24: Bingo!

1. Row 1 = 186
   Row 2 = 174
   Row 3 = 151
   Row 4 = 196
   Row 5 = 200
   Column B = 43
   Column I = 115
   Column N = 144
   Column G = 265
   Column O = 340
2. 49
3. 189
4. 16
5. 69

Page 25: Four-Digit Fun

1. 9,996
2. 9,753
3. 9,999
4. 8,888
5. 9,996
6. 9,900
7. 9,876
8. 9,984
9. 9,980
10. 9,801
11. 7,777
12. 9,000

Page 26: Counting Cookies

1. 40
2. 8
3. 31
4. 24
5. 9
6. 6
7. 6
8. 77
9. 72
10. 48

Page 27: Secret Code

Solution:

0 = m
1 = x
2 = t
3 = h
4 = c
5 = n
6 = d
7 = p
8 = z
9 = w

1. n
2. ht
3. tc
4. xh
5. t
6. xt
7. xn
8. cm
9. m
10. p

Page 28: A-"maze"-ing Pathways

1. 1-11-25-12-36-4-8-3-100
2. 100-8-12-27-3-6-26-17-1
3. 8-23-18-42-14-27-54-36-222

Page 29: Pick a Digit

1. 6,297
2. 2,621
3. 630
4. 416
5. 2,101,008
6. 2,579
7. 62,955

Pages 30–31: Shape Stumpers

1. 15
2. 10
3. 9
4. triangle
5. 6,545
6. 643
7. 195
8. 675
9. 323
10. 65
11. 255
12. 2,990

Page 32: Map Mysteries

1. 5,500 (Sacramento, Washington, 5, 11)
2. 1,287 (Massachusetts, Tallahassee, Tennessee)
3. 76 (Nebraska, Louisiana, Annapolis, Idaho)
4. 77 (Indianapolis, Salem, Missouri, Minnesota)
5. 960 (8 x 5 x 4 x 6)
6. 998 (10 x 8 x 12 + 32 + 6)

Page 33: Heads Up!

1. 18
2. 6
3. 9
4. 11
5. 18
6. 40

Page 34: Measure Up

1. 15,648 (15,840 − 432 + 240)
2. 128 (40 + 56 + 32)
3. 738 (18 + 720)
4. 5,890 (4,320 + 840 + 730)
5. 392 (312 + 140 − 60)
6. 12,600 (600 + 6,000 + 6,000)
7. 4,966 (880 + 126 + 3,960)
8. 772 (72 + 70 + 630)

Page 35: Five-Digit Fun

1. 77,777
2. 99,999
3. 99,997
4. 99,000
5. 99,500
6. 99,975
7. 99,600
8. 98,765
9. 90,009
10. 99,856
11. 95,210
12. 97,532

Page 36: Flowing Along
1. 4,824
2. 15,264
3. 25,488
4. 36,144
5. 43,272

Page 37: Fill 'Er Up

Fill-ins:
Row 1: 25, 60
Row 2: 62
Row 3: 3, 47
Row 4: 73
Row 5: 28

Totals:
Row 1 = 200
Row 2 = 175
Row 3 = 146
Row 4 = 216
Row 5 = 183
Column B = 43
Column I = 122
Column N = 157
Column G = 258
Column O = 340

Page 38: Make a Date
1. May 18
2. April 27
3. June 3
4. June 11
5. May 27
6. March 11
7. February 28
8. Friday
9. Tuesday
10. Wednesday
11. Monday
12. Friday

Page 39: Something in Common

Examples:
1. 51 (First digit is 4 more than second digit)
2. 808 (all are palindromes)
3. 89 (each digit one more than previous)
4. 422 (sum of digits equals 8)
5. 68 (each digit 2 more than previous)
6. 53 (each digit is prime)
7. 134 (middle digit is difference between first and third digits)

Page 40: Almanac Math
1. 3,625 (1959, 1803, 62, 75)
2. 450 (5, 5, 9, 2)
3. 7,330 (1751, 1829, 1885, 1865)
4. 2,298 (555, 1922, 1814, 1993)
5. 3,941 (1943, 1950, 1989, 1941)
6. 3,812 (1932, 1904, 1936, 1960)
7. 3,711 (1817, 1885, 1982, 1973)

Page 41: Three-Sided Puzzles
1. 10 (A + B − C)
2. 8 (A × B ÷ C)
3. 8 (A ÷ B + C)
4. 9 (C − A + B)
5. 13 (A × C − B)

Page 42: Keypad Challenge

Examples:
1. cabin, bacon
2. backed, cabled
3. black, vault
4. seed, reef
5. hog/log, Kim/him
6. labels, packed
7. white, snowy
8. waxy, quit
9. high, noon, feed
10. cable, ahead

Page 43: Alphabet Alley

Examples:
1. faded
2. cages
3. decal
4. woozy
5. geeks
6. faced
7. color
8. moons
9. grabs
10. apple

Page 44: Playing with Palindromes

1. 112
2. 224
3. 336
4. 505
5. 151
6. 717
7. 393
8. 929
9. 1,460
10. 1,010
11. 99
12. 222
13. 2,112
14. 2,002
15. 8,998

Page 45: Probably Not!

1. 3:4
2. 7:8
3. 1:2

1. 4:13
2. 10:13
3. 8:13

1. 1:2
2. 1:4
3. 5:12
4. 1:2
5. 7:12

Pages 46: Math Mysteries

1. 3:55
2. $6.95
3. 40 students
4. $16.00
5. Abigail: $1.20
   Ben: $2.20
   Chaz: $.60
   Drew: $2.40

Page 47: Tons of Numbers

1. 559
2. 1,048
3. 776
4. 384
5. 279
6. 448

Page 48: Prime Time

1. 28
2. 41
3. 679
4. 27
5. 120
6. 107
7. 269
8. 2
9. 86
10. 4,757
11. 264
12. 3,127

Page 49–50: Prize Patrol

LIC. PLATE	MODEL	NAME
123–HTT	Subaru	Lynn Talbert
573–BHD	Dodge	Kara Stenson
108–CXL	Lexus	Harry Pendleton
111–YLB	Mitsubishi	Nora Bland
646–MZA	Kia	Scott Alderson
392–JGJ	Honda	Steve Marsh
416–PLK	Nissan	Meg Jarvis
377–AQW	Saturn	Derek Freeman

Page 51: Calculator Products

1. least: 38 x 26
   greatest: 63 x 82
2. least: 69 x 68
   greatest: 96 x 86
3. least: 58 x 27
   greatest: 82 x 75
4. least: 69 x 27
   greatest: 92 x 76
5. least: 26 x 479
   greatest: 762 x 94
6. least: 27 x 389
   greatest: 872 x 93

Page 52: Go Figure

1. 291
2. 1,582
3. 2,364
4. 8,653

Page 53: Exploring the Americas

1. 77
2. 40
3. 33
4. 28
5. 63
6. 14,493

Page 54: Feline Fun

345–623 = Fuzzy
424–355 = Fifi
611–253 = Fred
608–331 = Fluffy
497–209 = Fillis
777–902 = Flair
691–224 = Finicky
139–267 = Flash

Page 55: Don't Be Square

1. 108
2. 56
3. 72
4. 21.5
5. 250
6. 64
7. 204
8. 12

Page 56: A–B–C Math

Examples:

1. A dog ate a bone.
2. hippopotamus
3. answers will vary
4. arrests
5. bear
6. a. Ontario (92 cents)
   b. Antarctica (90 cents)
   c. Atlantic (80 cents)
   d. November (94 cents)
   e. Tuesday (95 cents)
   f. California (88 cents)

Page 57: It's a Dog's Life

CBT–841 = Benji
ACT–212 = Bernice
XRT–567 = Bojo
YWE–118 = Buzzy
SSH–362 = Beast
KLP–136 = Brownie
LJD–407 = Black Jack
NNO–782 = Bart